Dig.

Paleontologists dig for fossils.

Dig.

Dig. Lift. Sift.

Sometimes man-made objects are found.

Let’s dig some more. What will we find?

Scoop.

Sand dollars are a type of sea urchin.

What's under all these rocks?

Heave.

No dinosaur yet.

Plesiosaurus was a reptile that lived in the ocean.

Dig.

Chip, chip, chip.

Coprolites are fossilized poop.

Dig.

Are we getting close?

ROCKSLIDE!

Rockslides on coastal cliffs can uncover fossils, stones, and bones.

Scoop. Lift.

Push.

Scientists have to deal with weather, nature, and animals on dig sites.

We need a bigger shovel.